Saab J 35 Draken

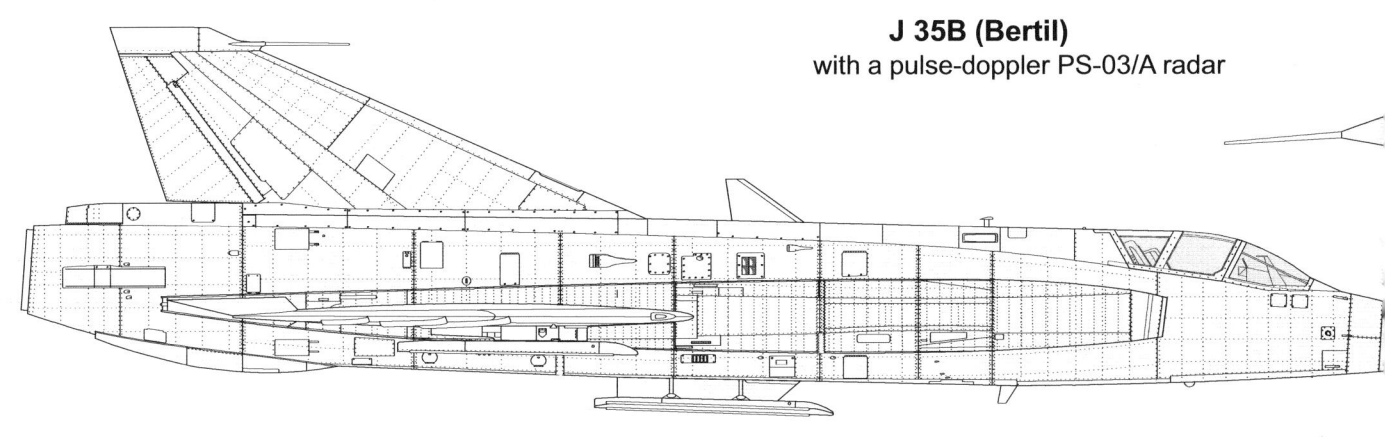

J 35B (Bertil)
with a pulse-doppler PS-03/A radar

J 35BS
J 35B version for Finland

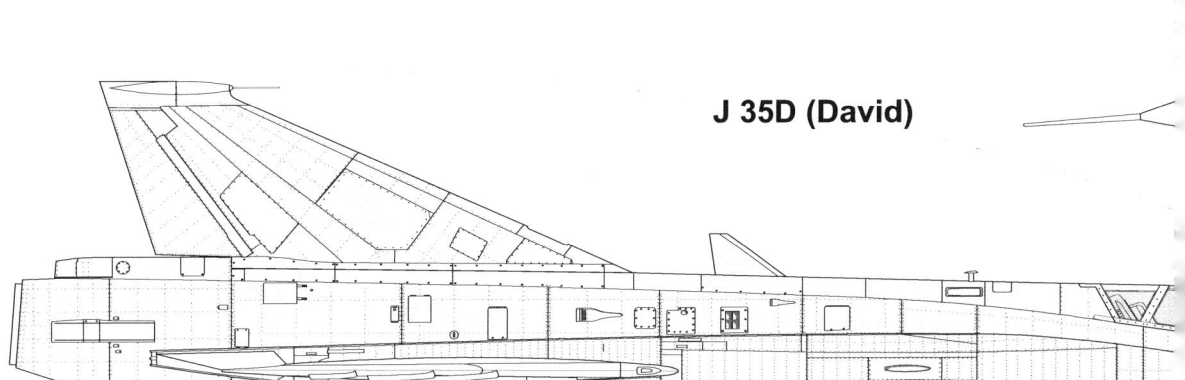

J 35D (David)

Saab J 35 Draken

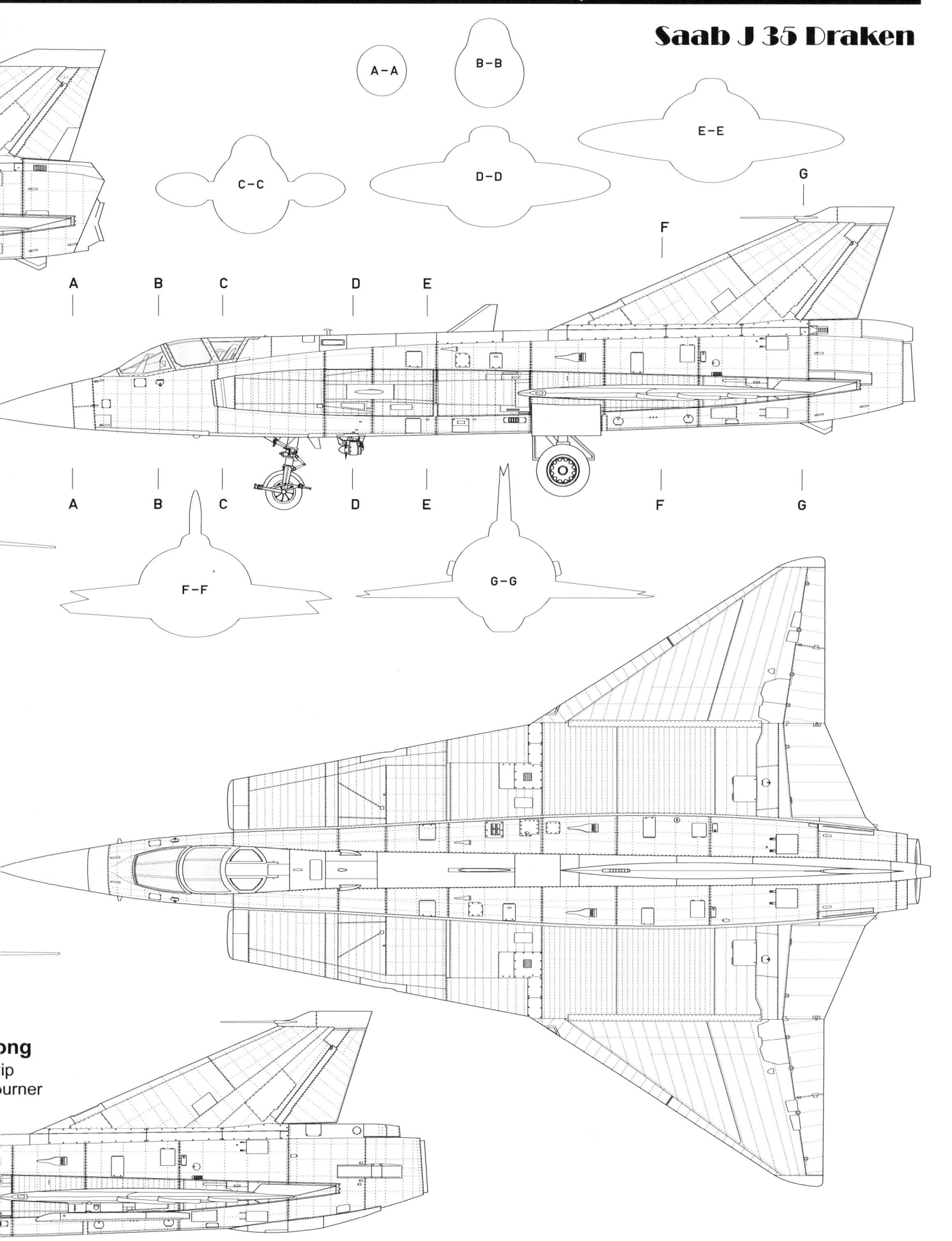

A–A

B–B

E–E

C–C

D–D

G

F

A B C D E

A B C D E F G

F–F

G–G

ong

ip

urner

1

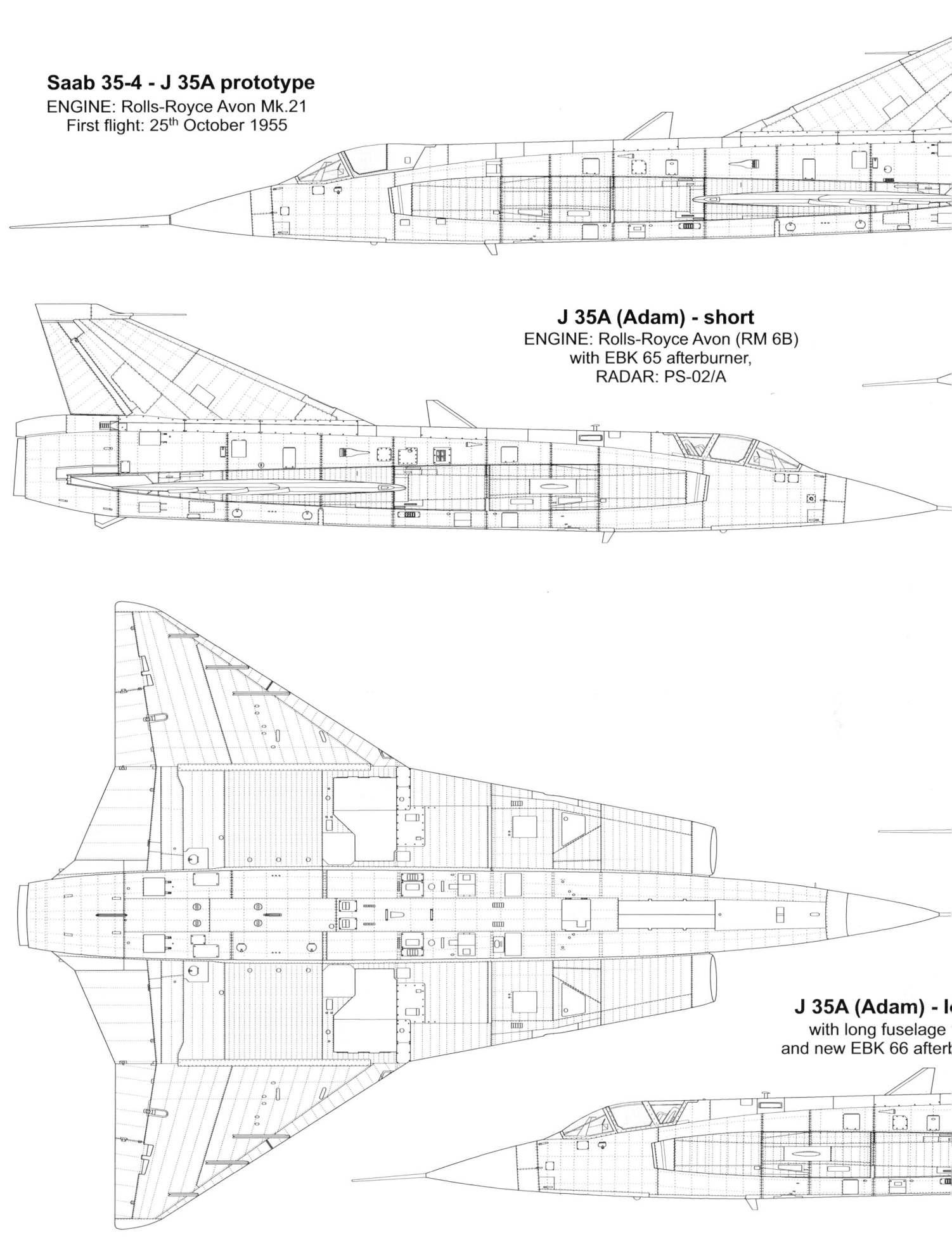

Saab 35-4 - J 35A prototype
ENGINE: Rolls-Royce Avon Mk.21
First flight: 25th October 1955

J 35A (Adam) - short
ENGINE: Rolls-Royce Avon (RM 6B)
with EBK 65 afterburner,
RADAR: PS-02/A

J 35A (Adam) - l
with long fuselage
and new EBK 66 afterb

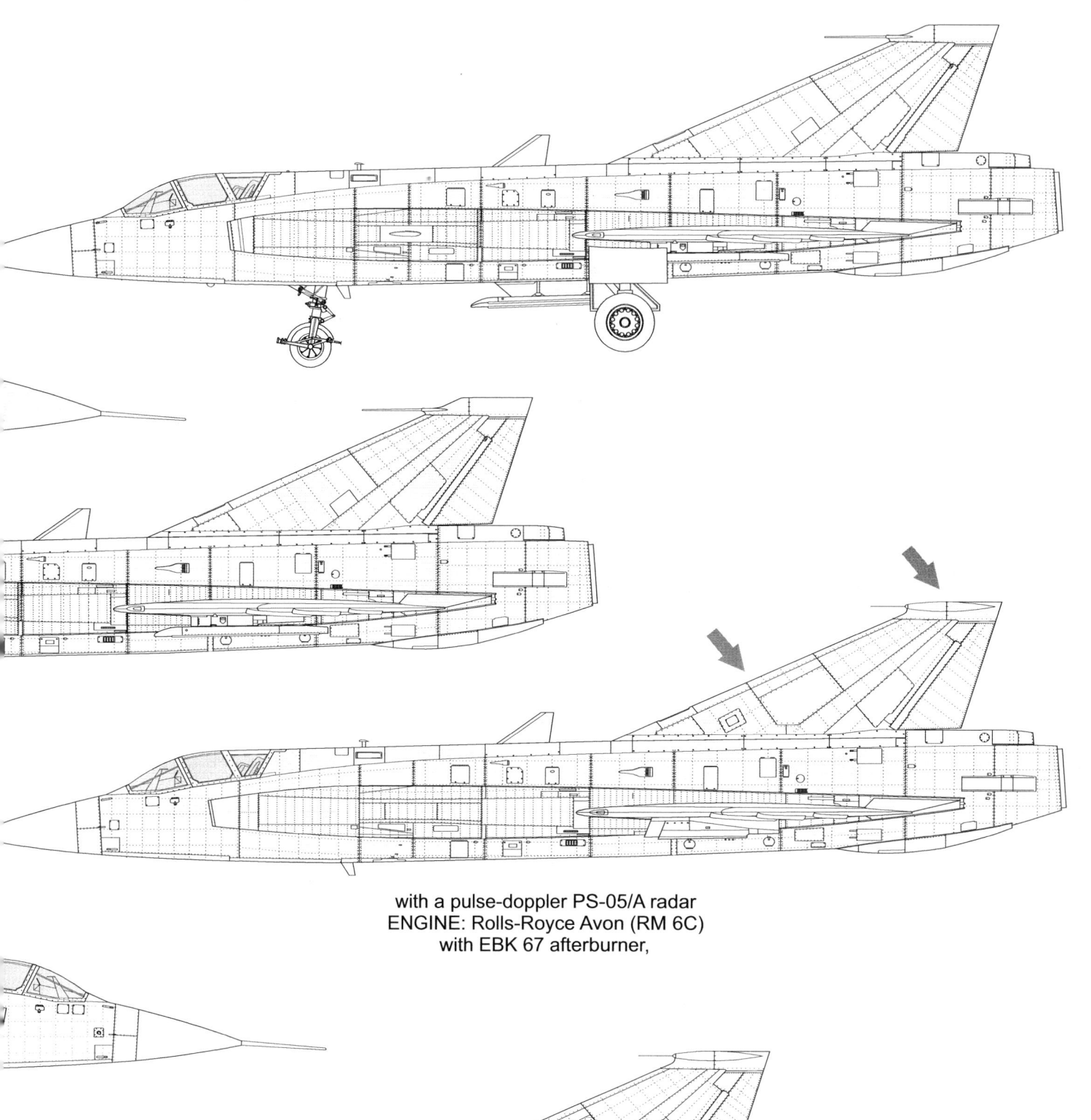

with a pulse-doppler PS-05/A radar
ENGINE: Rolls-Royce Avon (RM 6C)
with EBK 67 afterburner,

J 35OE
J 35D version for Austria

Saab J 35 Draken

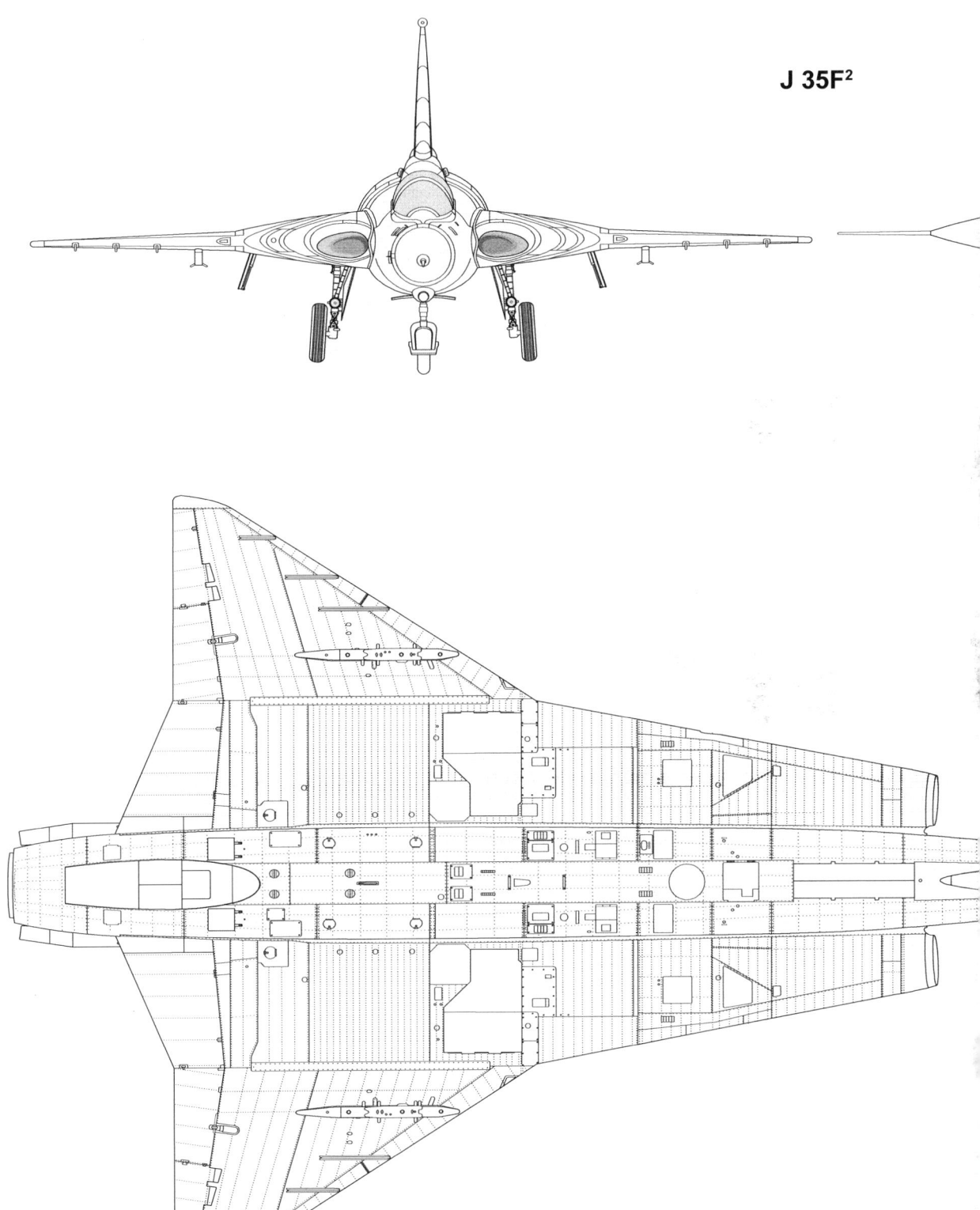

J 35F²

Saab J 35 Draken

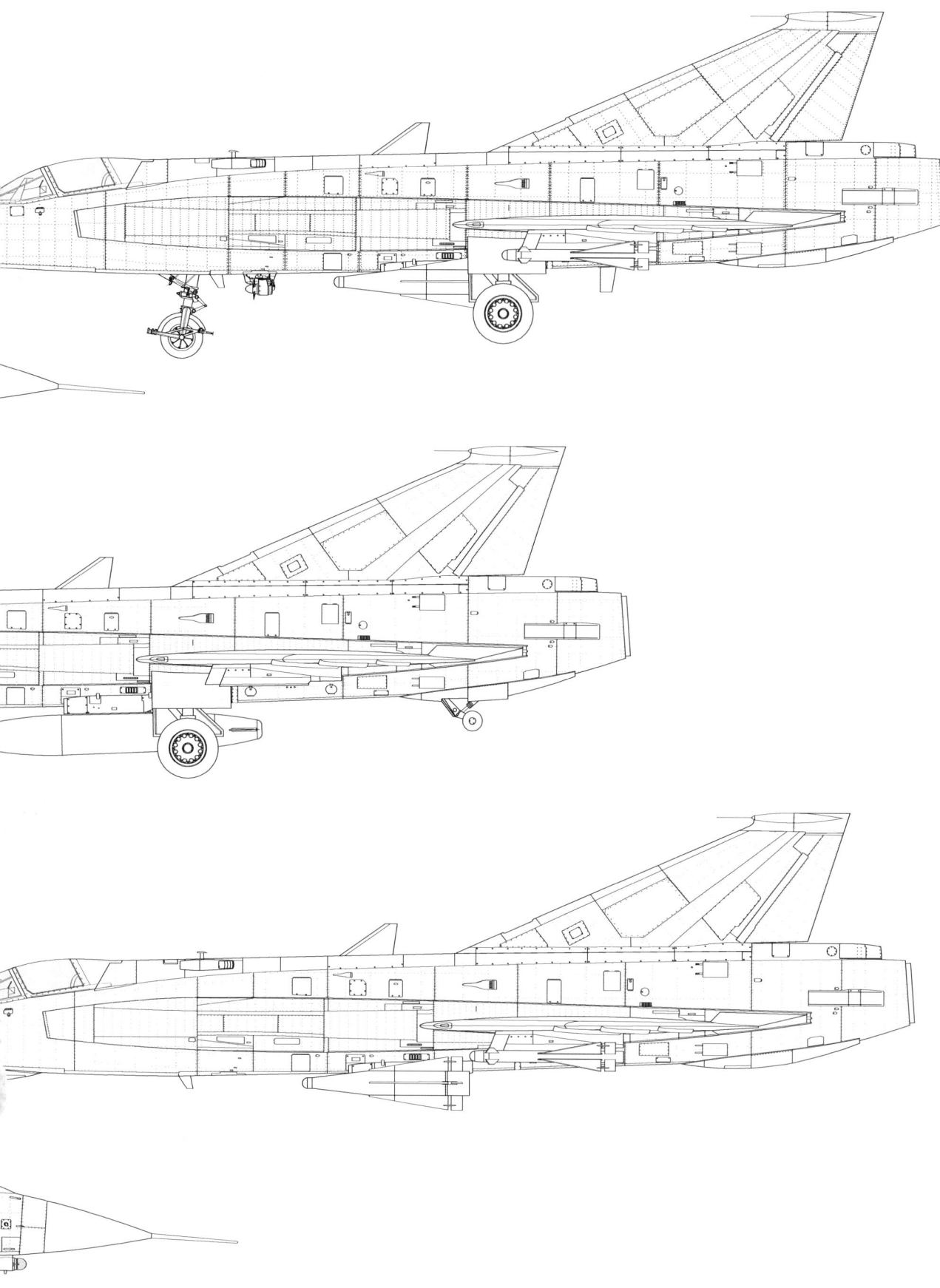

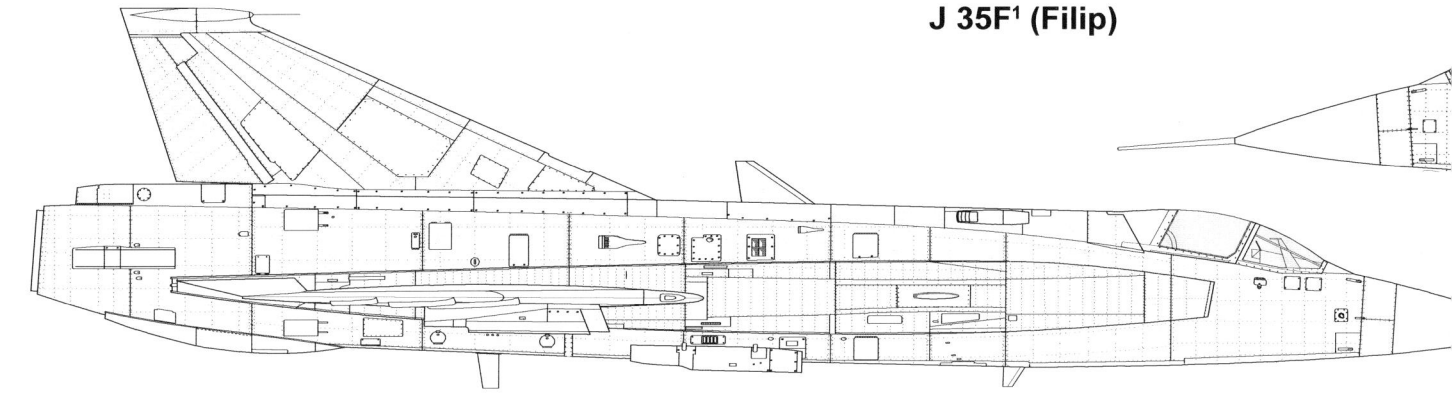

J 35F¹ (Filip)

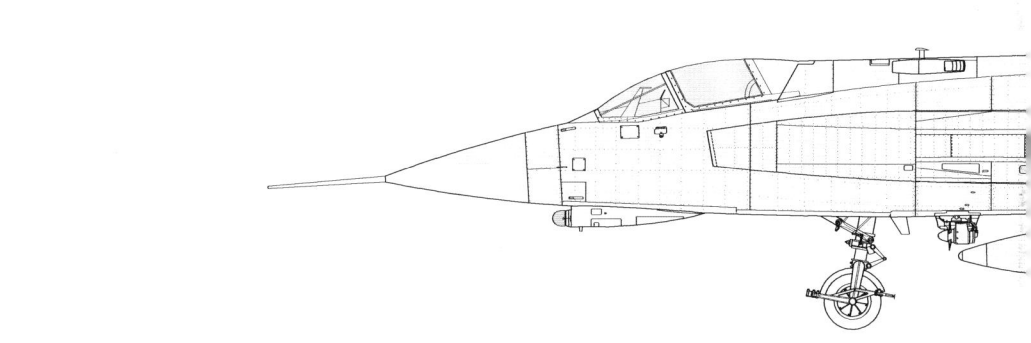

J 35F² (Filip)

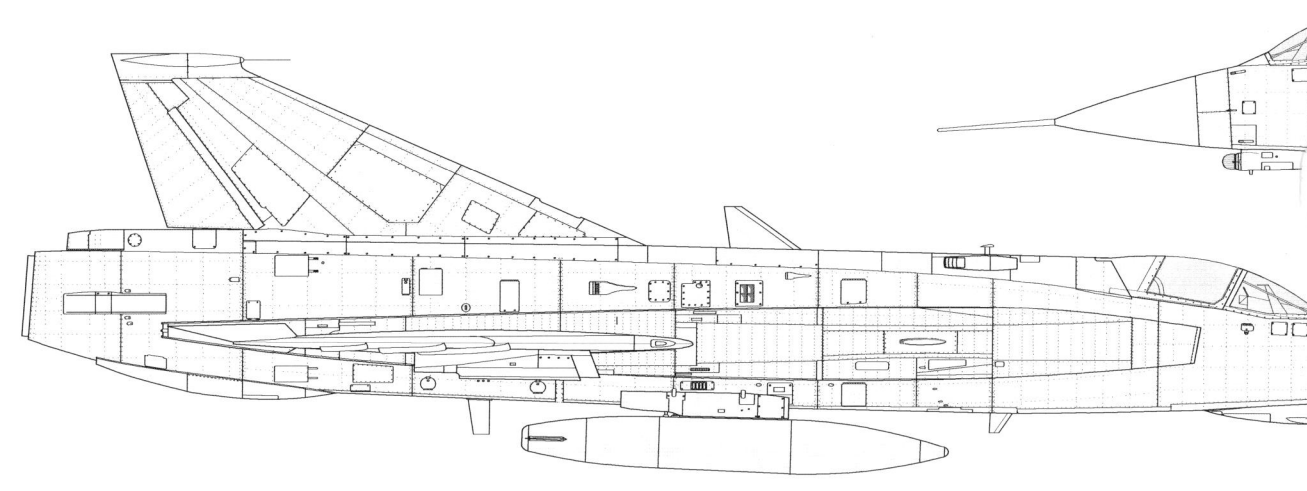

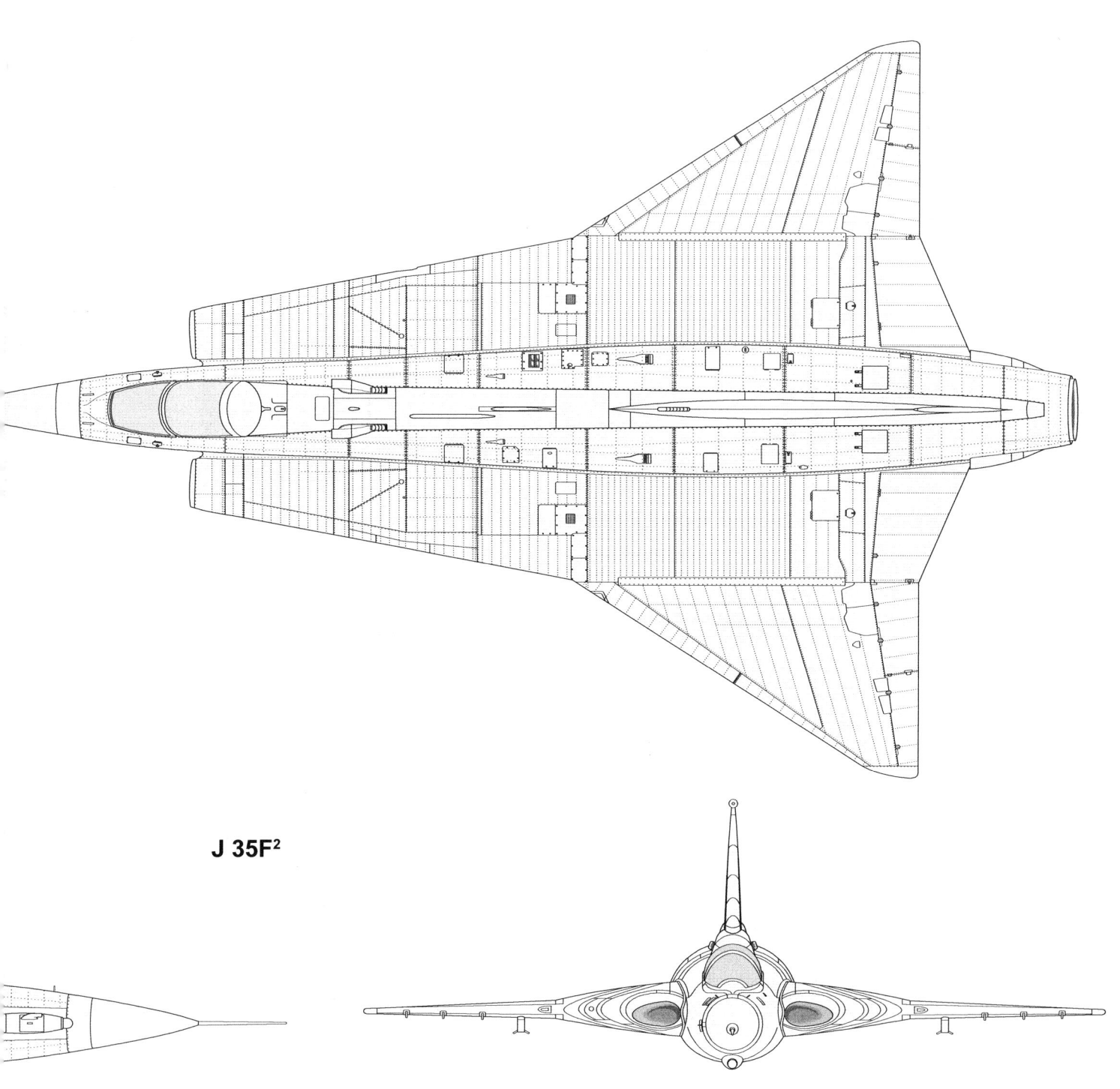

J 35F²

Saab J 35 Draken

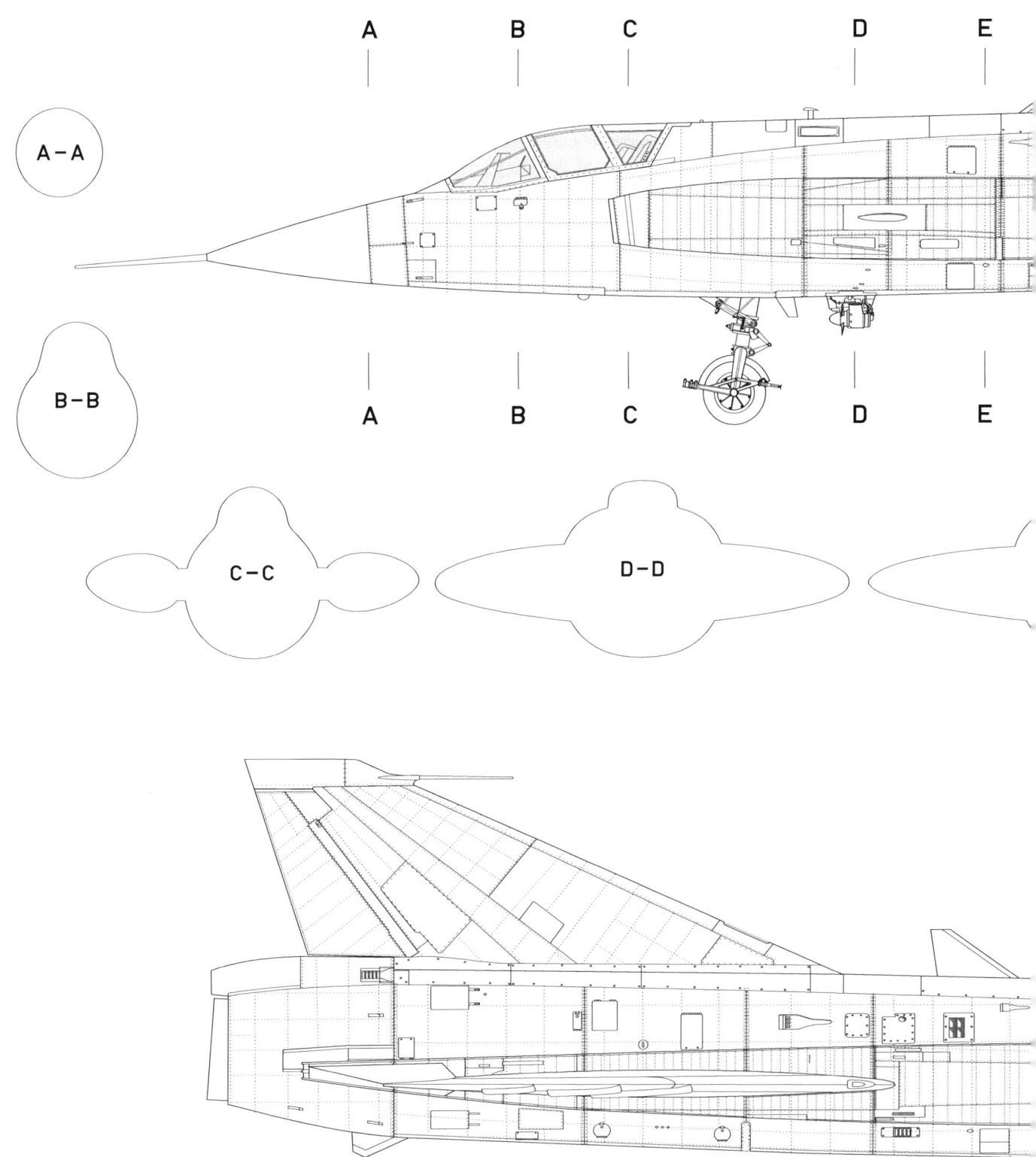

A–A

B–B

C–C

D–D

Saab J 35 Draken

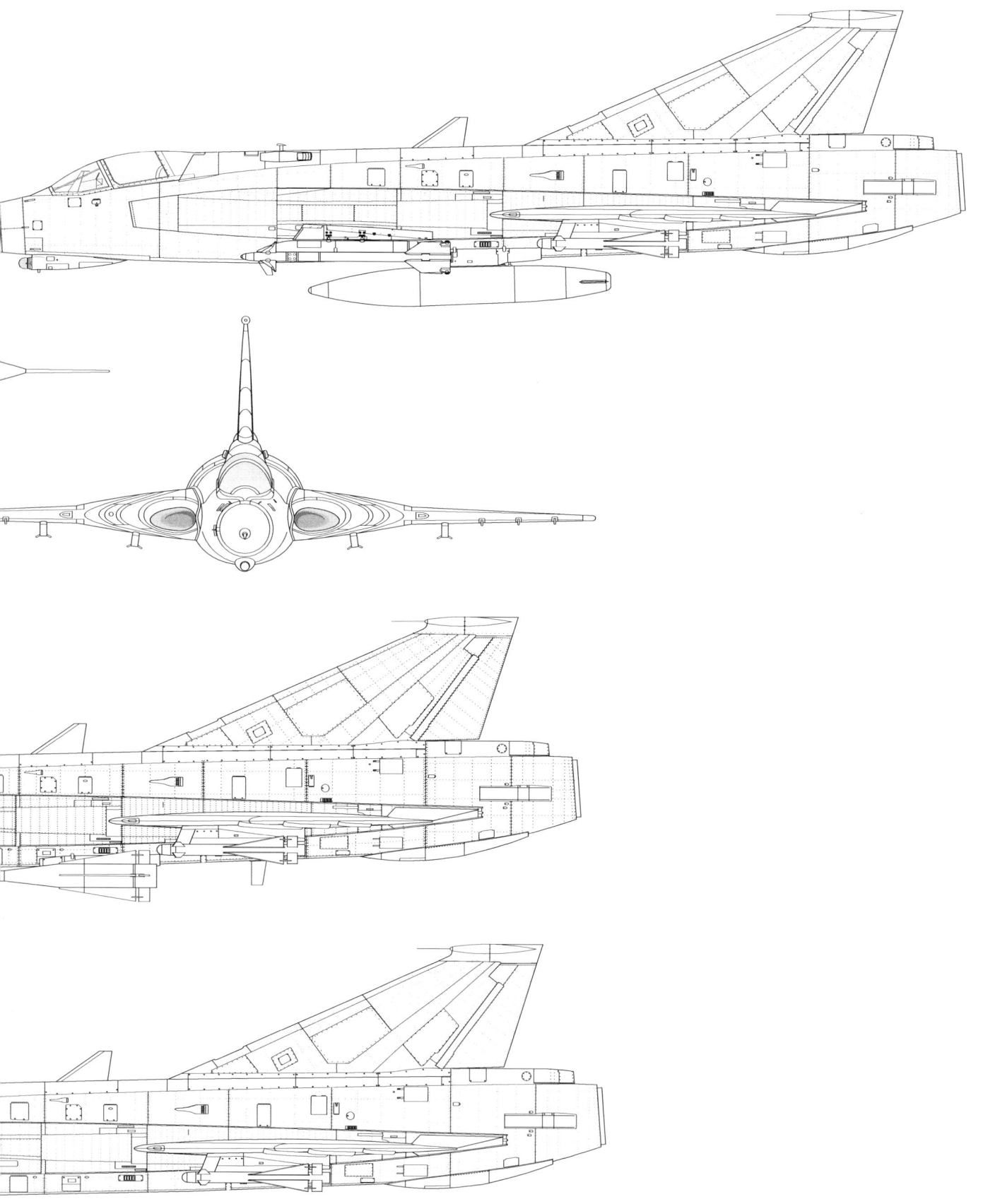

J 35J (Johan)

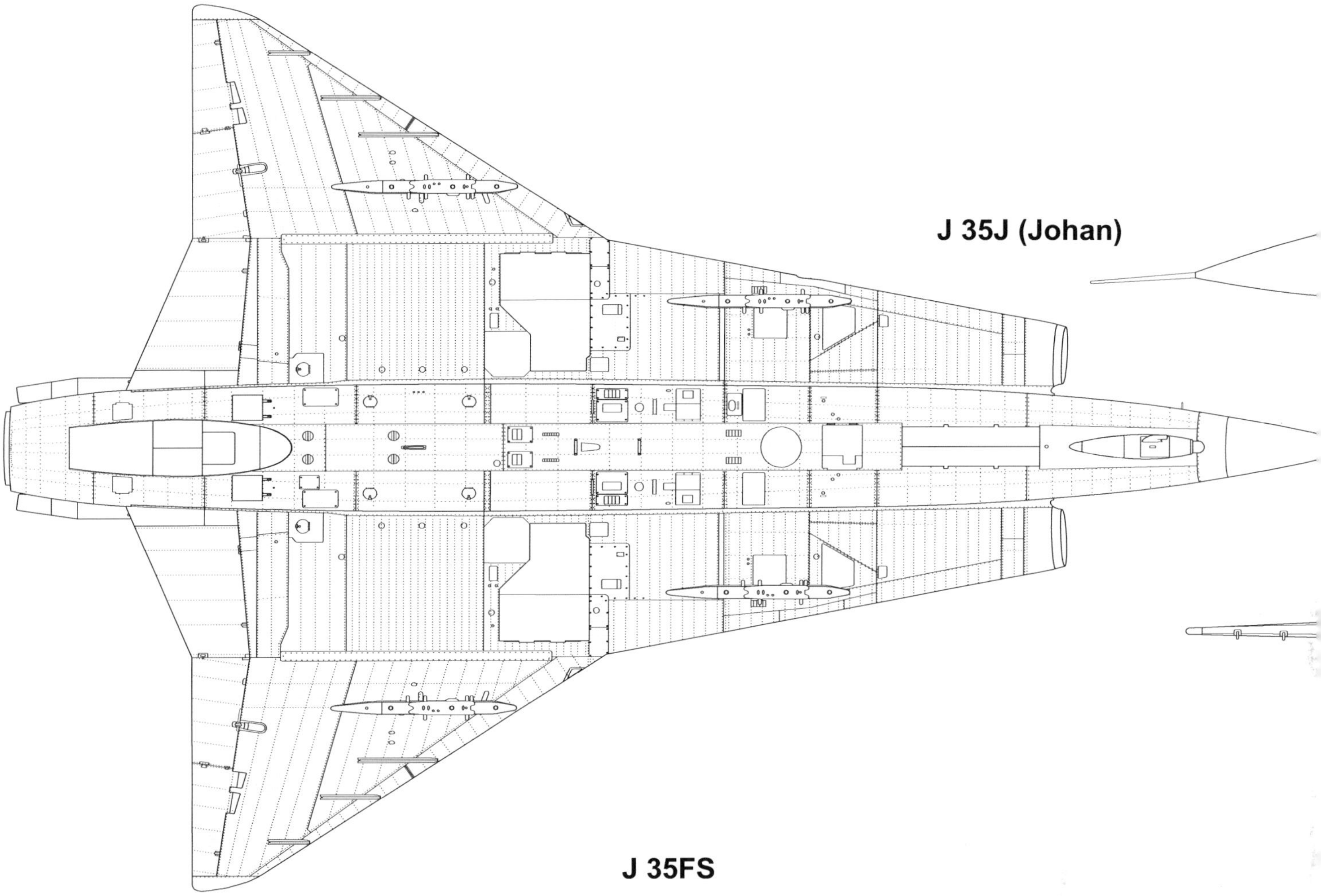

J 35FS

J 35F[1] version for Finland

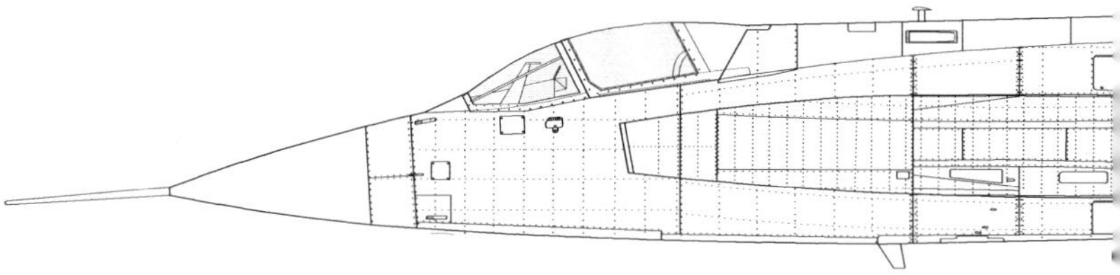

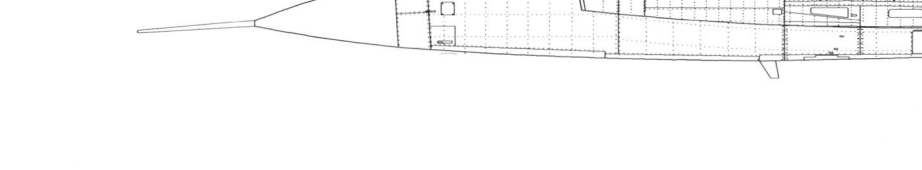

J 35S

J 35F[2] version
"manufactured" in Finland for Finland

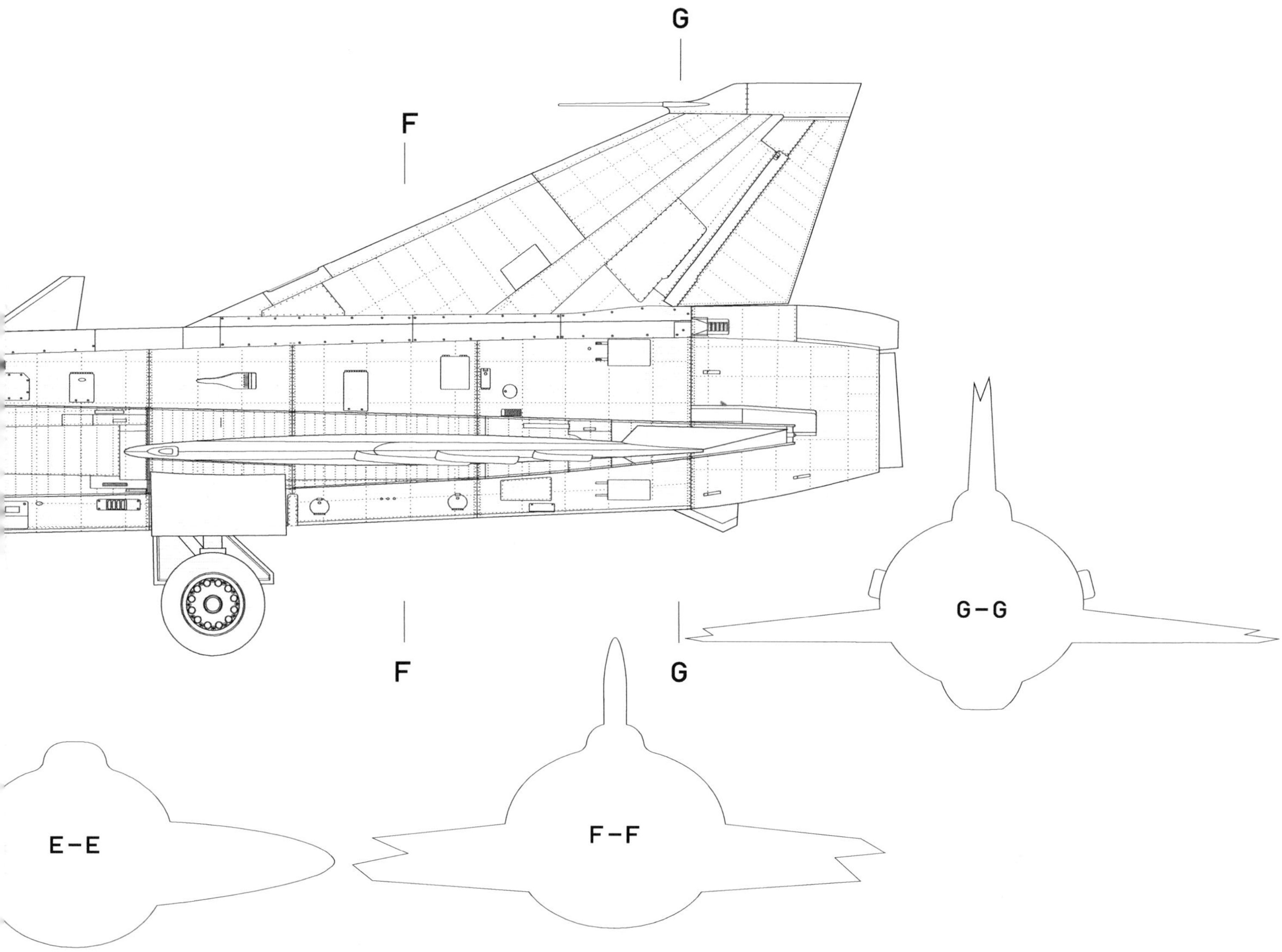

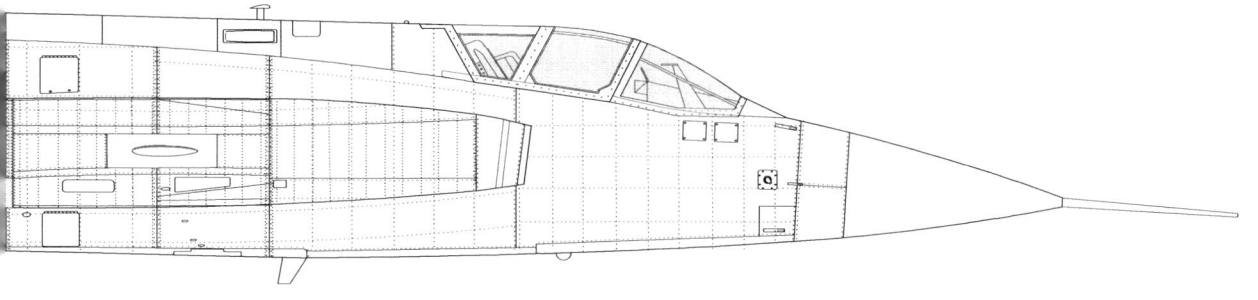

G

F

F

G

G–G

F–F

E–E

J 35A

Saab J 35 Draken

J 35B

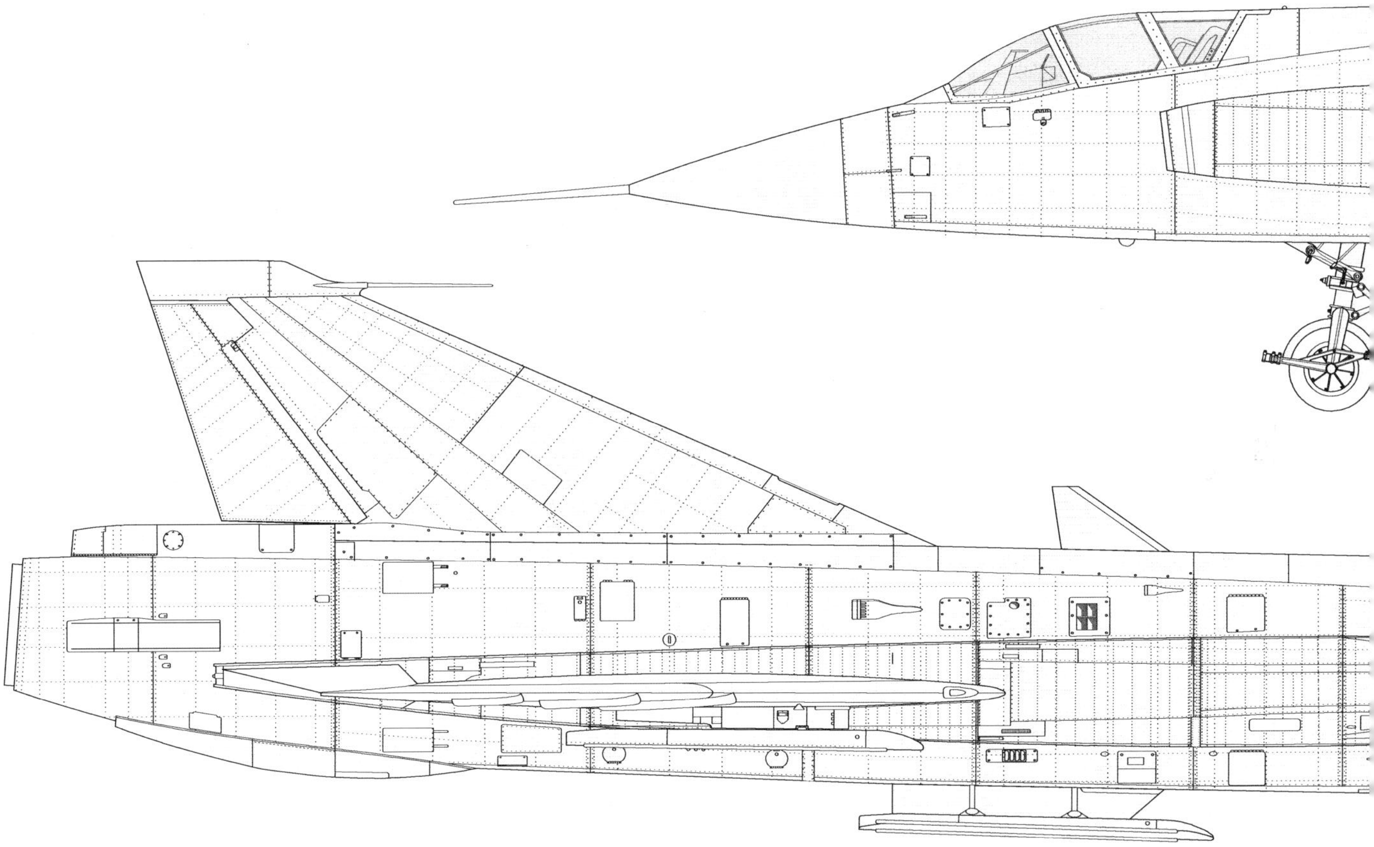

J 35D

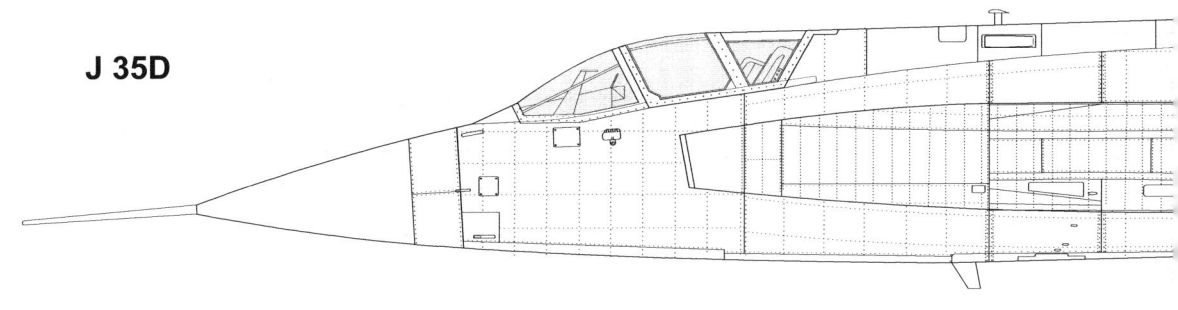

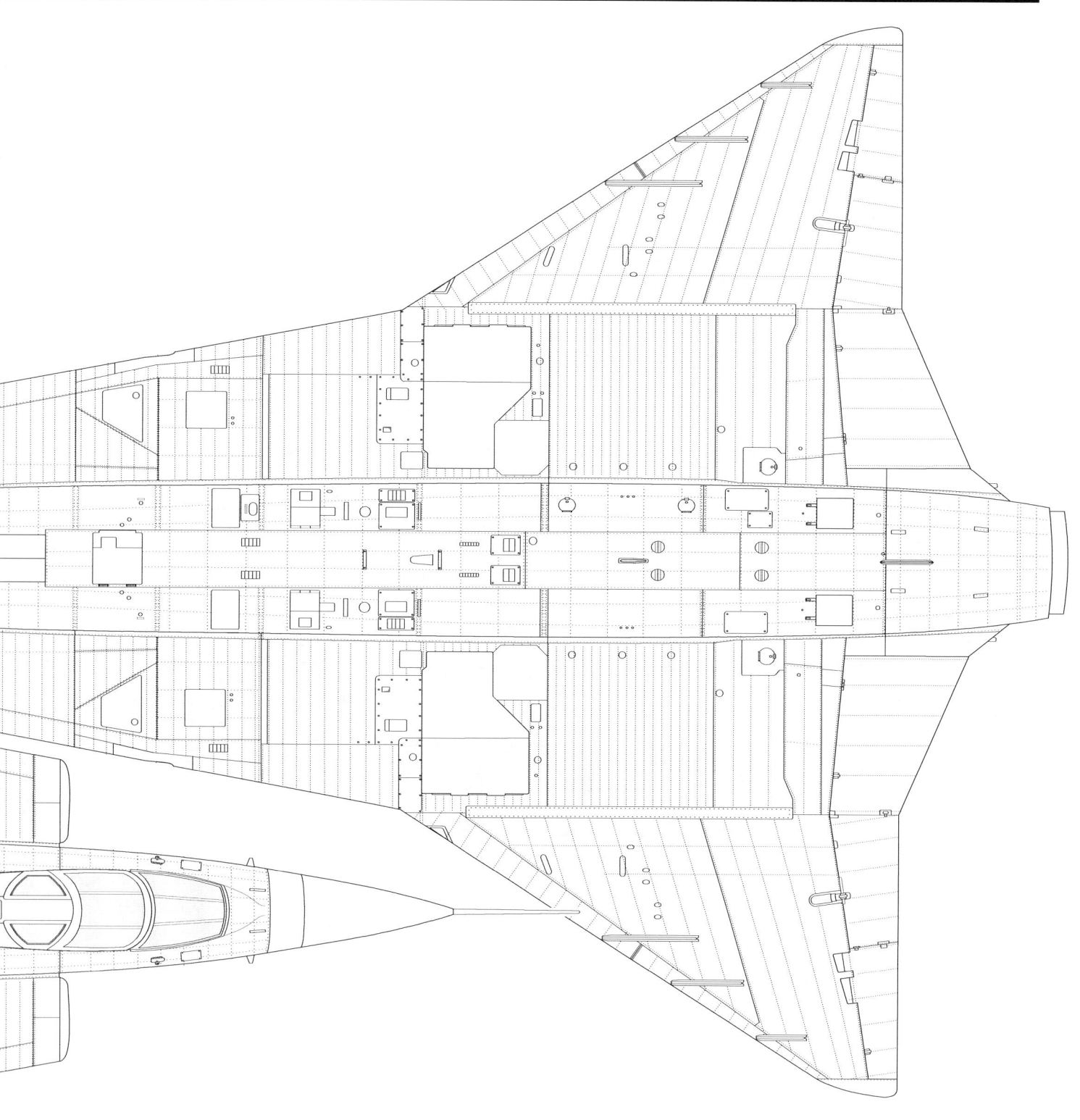

7

Saab J 35 Draken

J 35 A

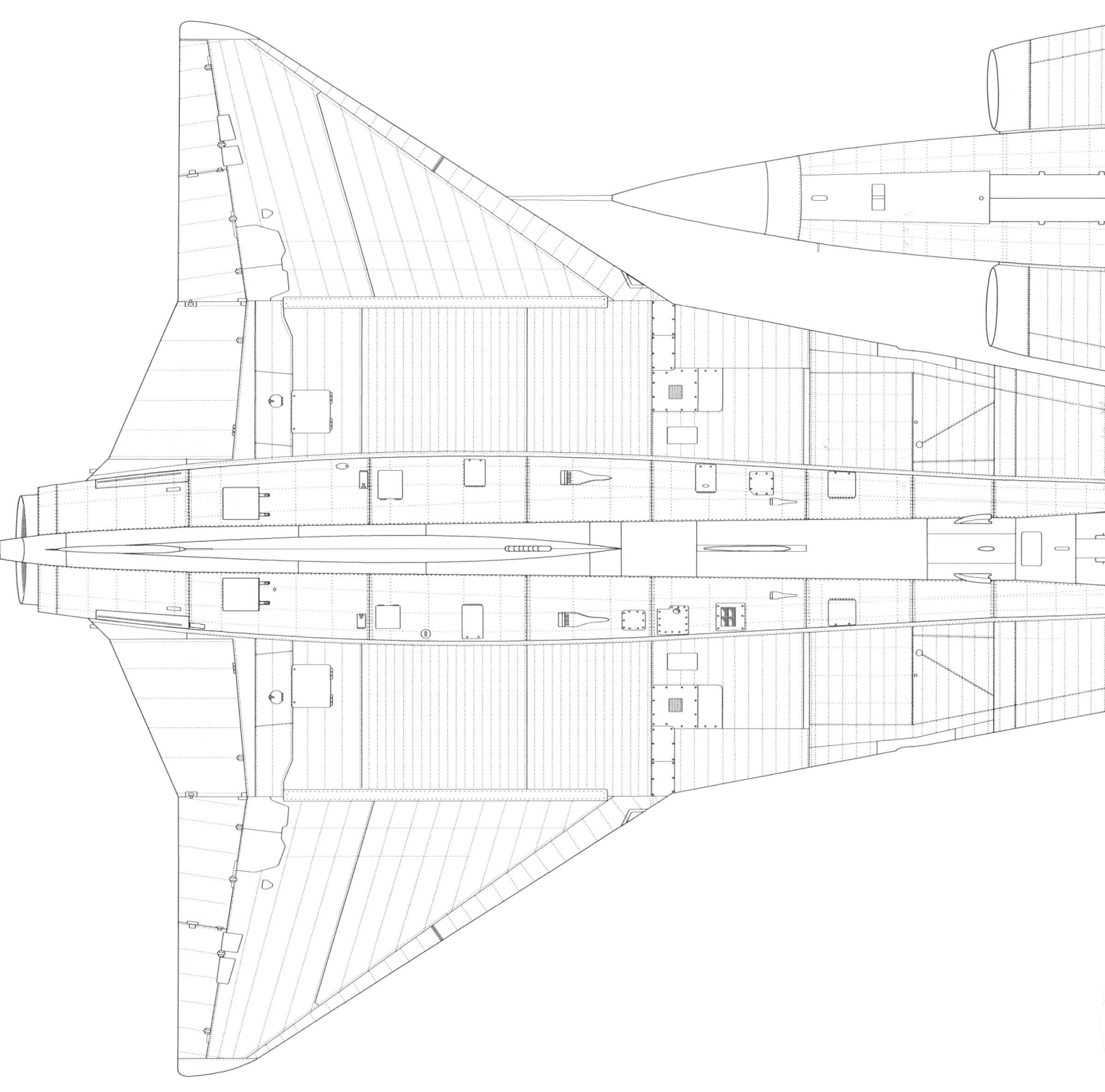

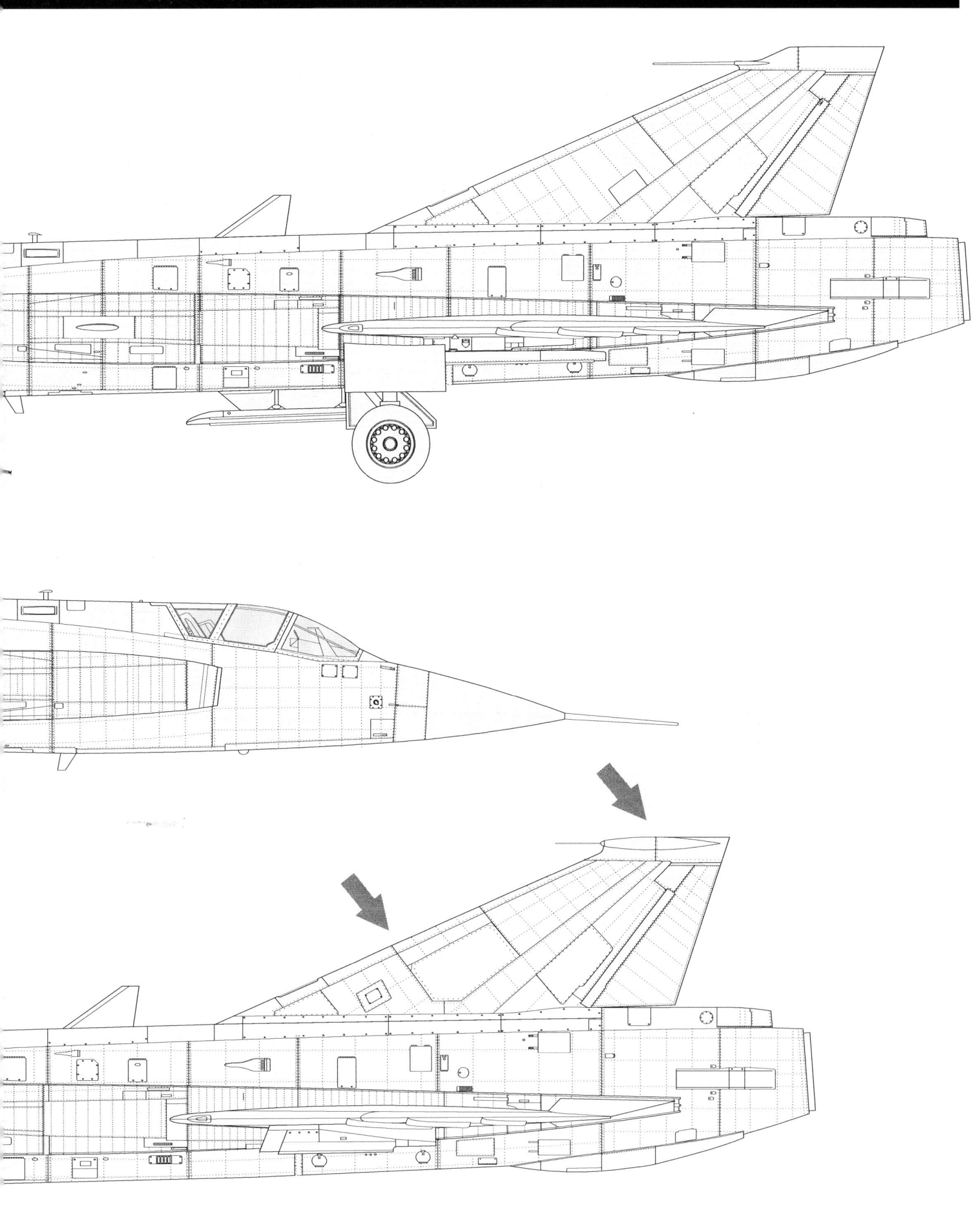

Saab J 35 Draken

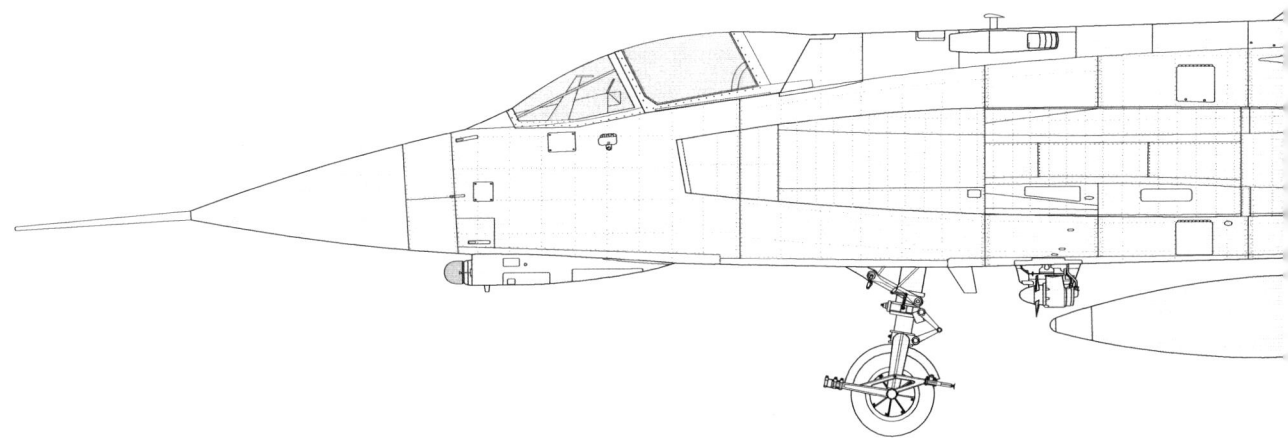

J 35F²

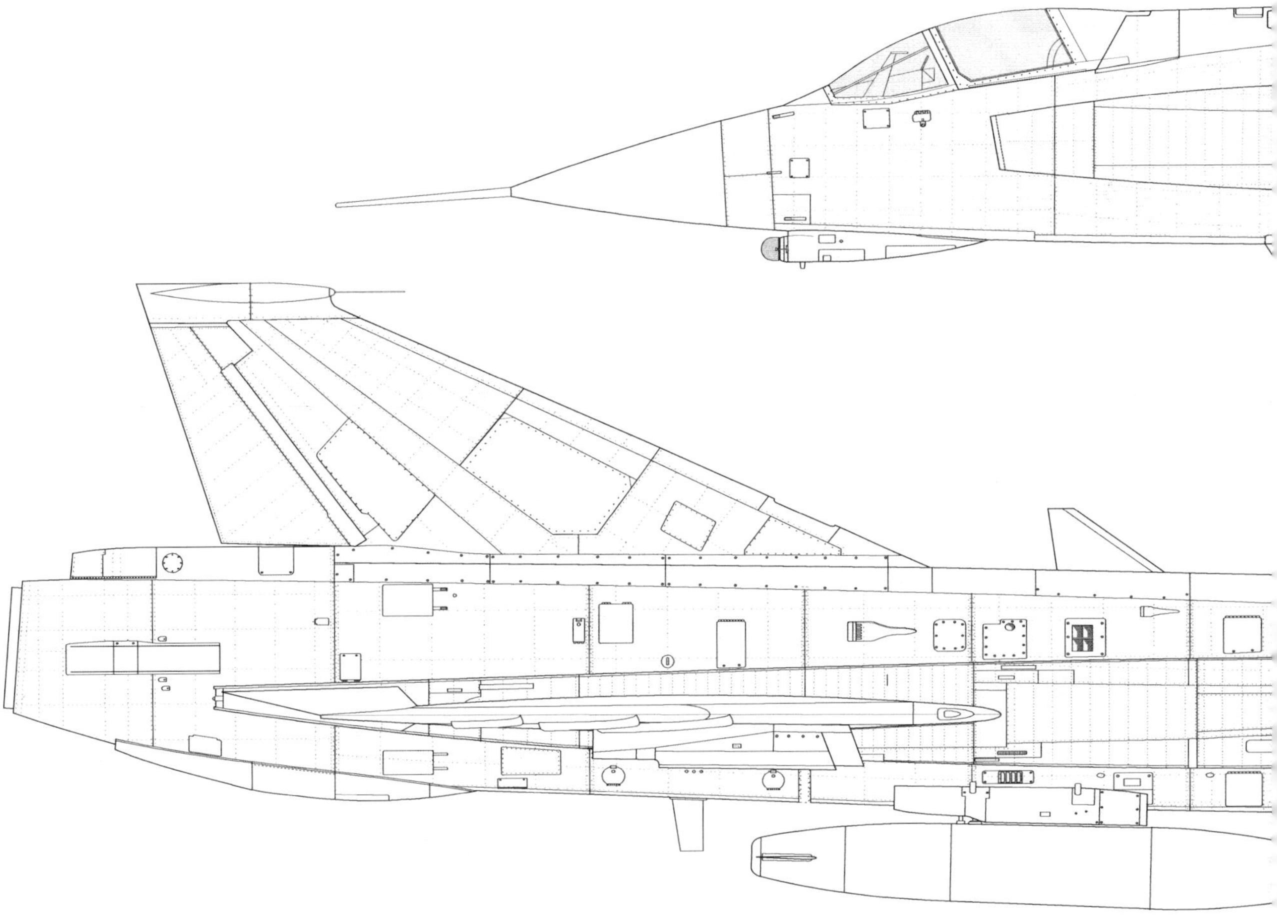

Saab J 35 Draken

J 35D

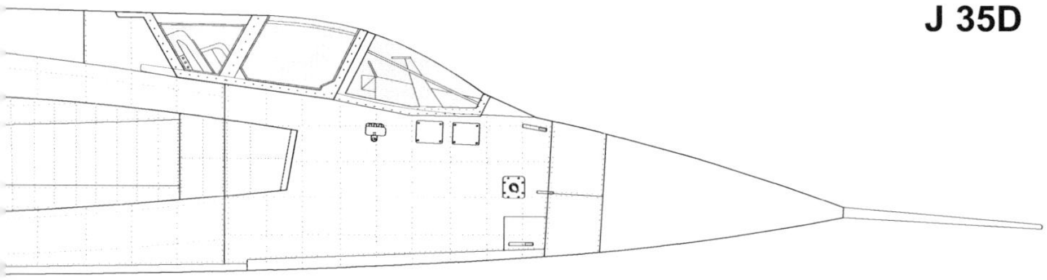

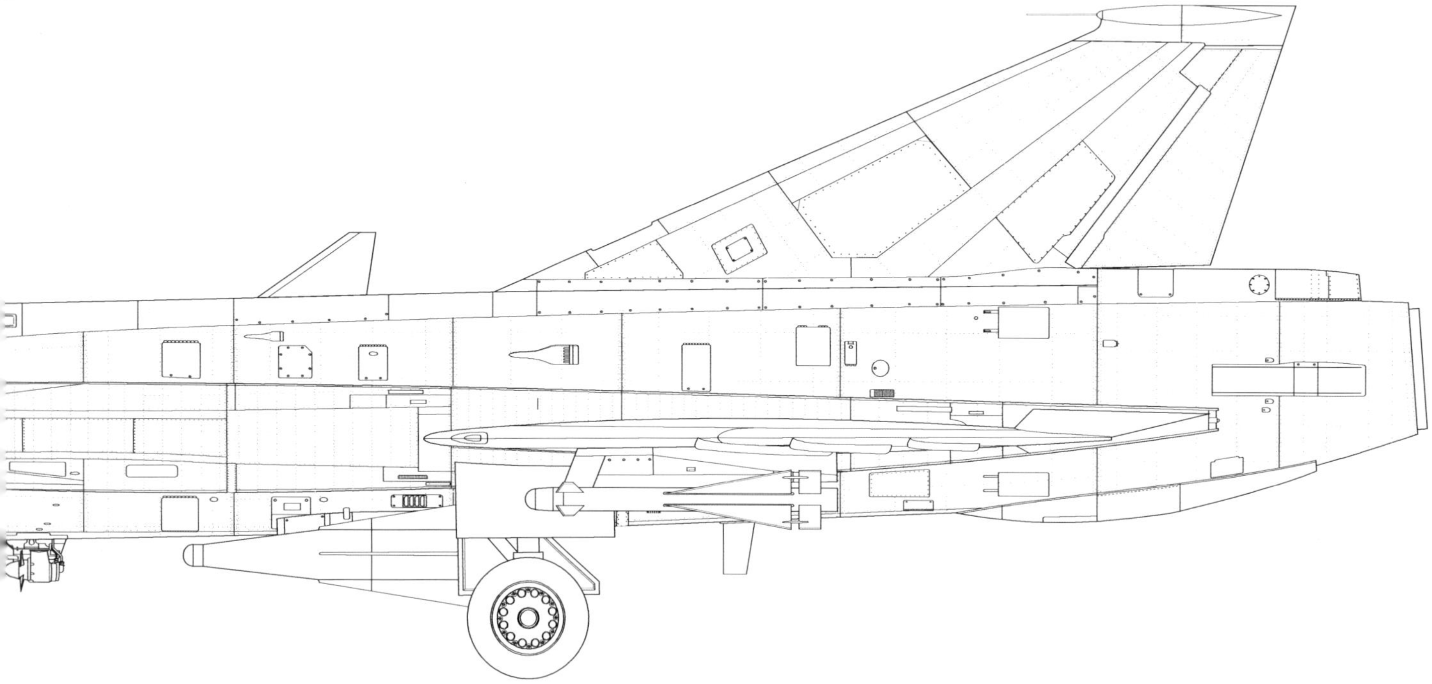

J 35F[1]

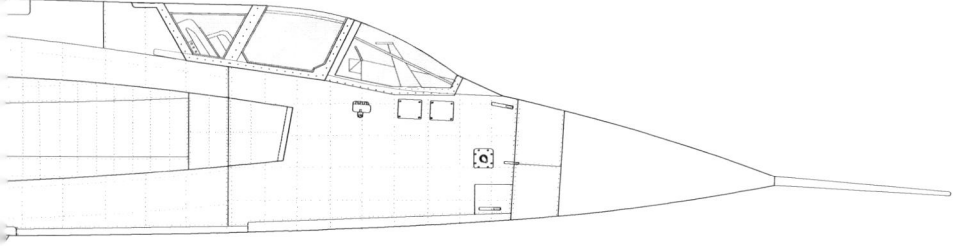

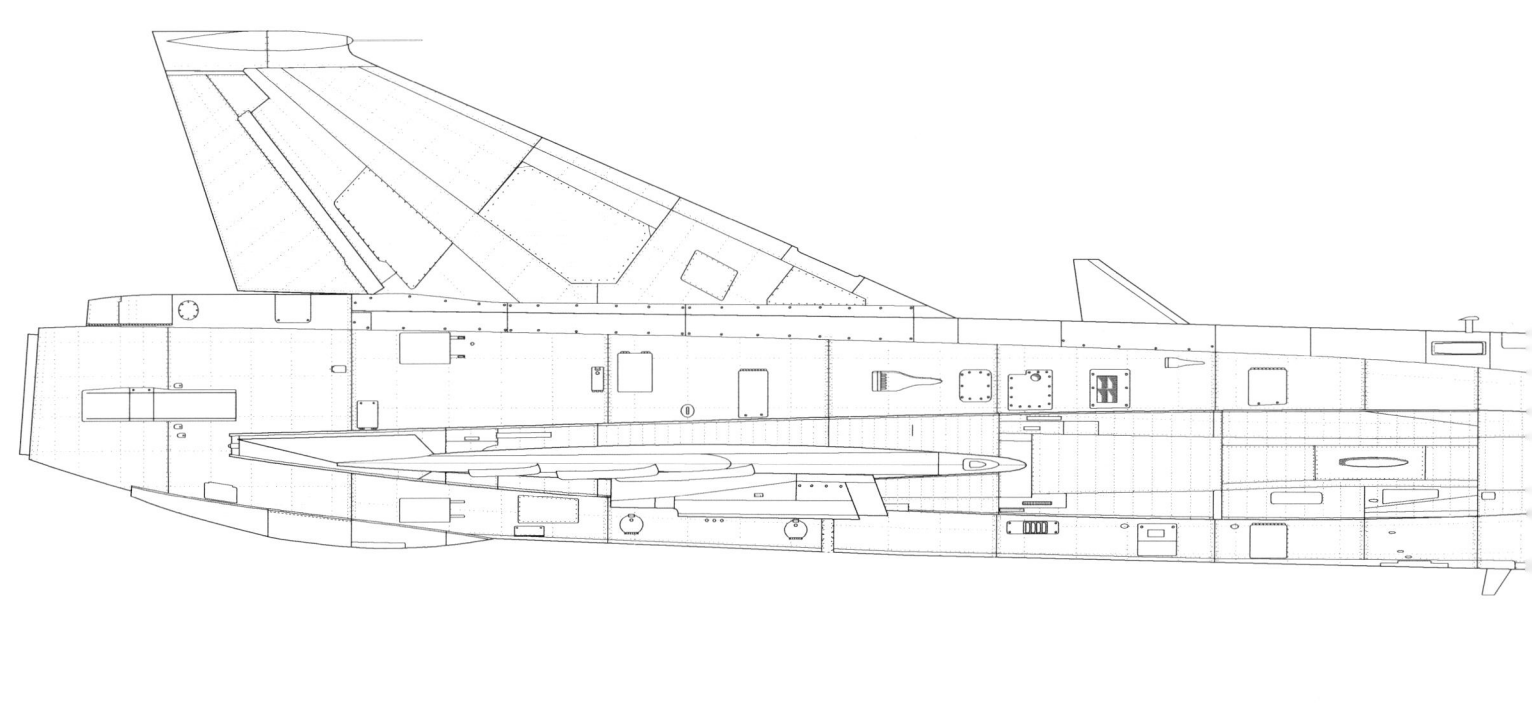

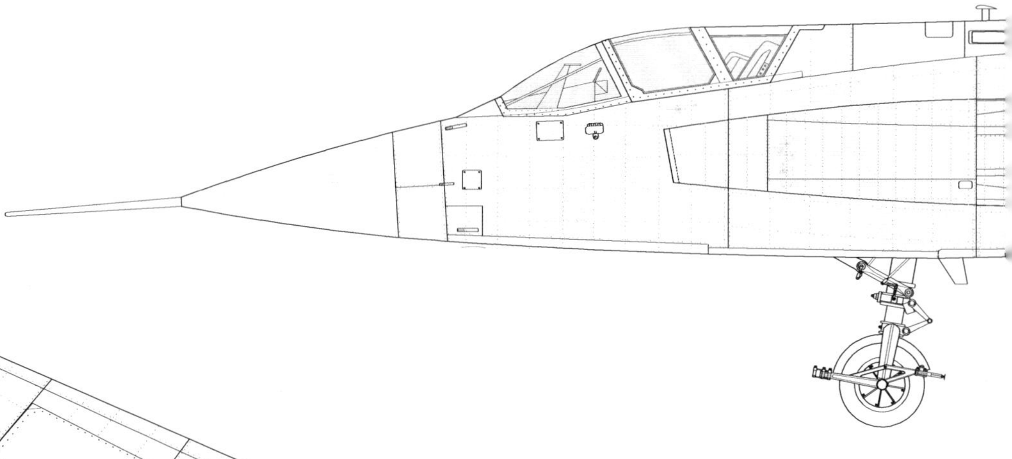

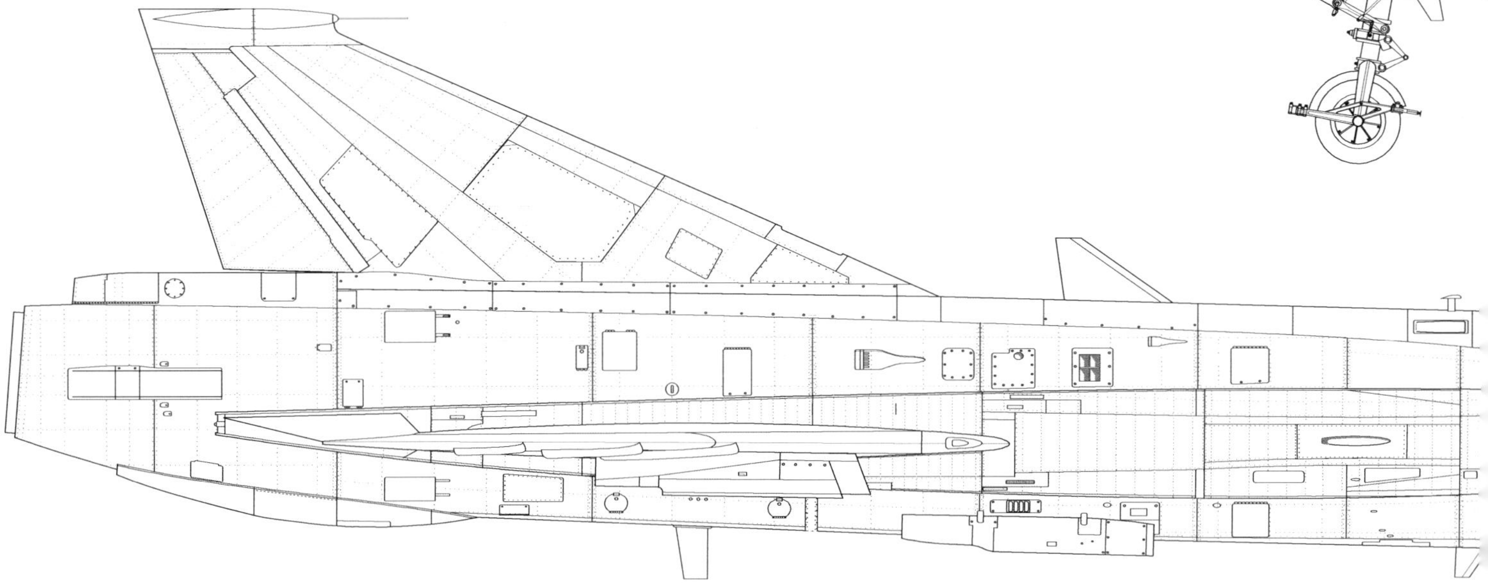

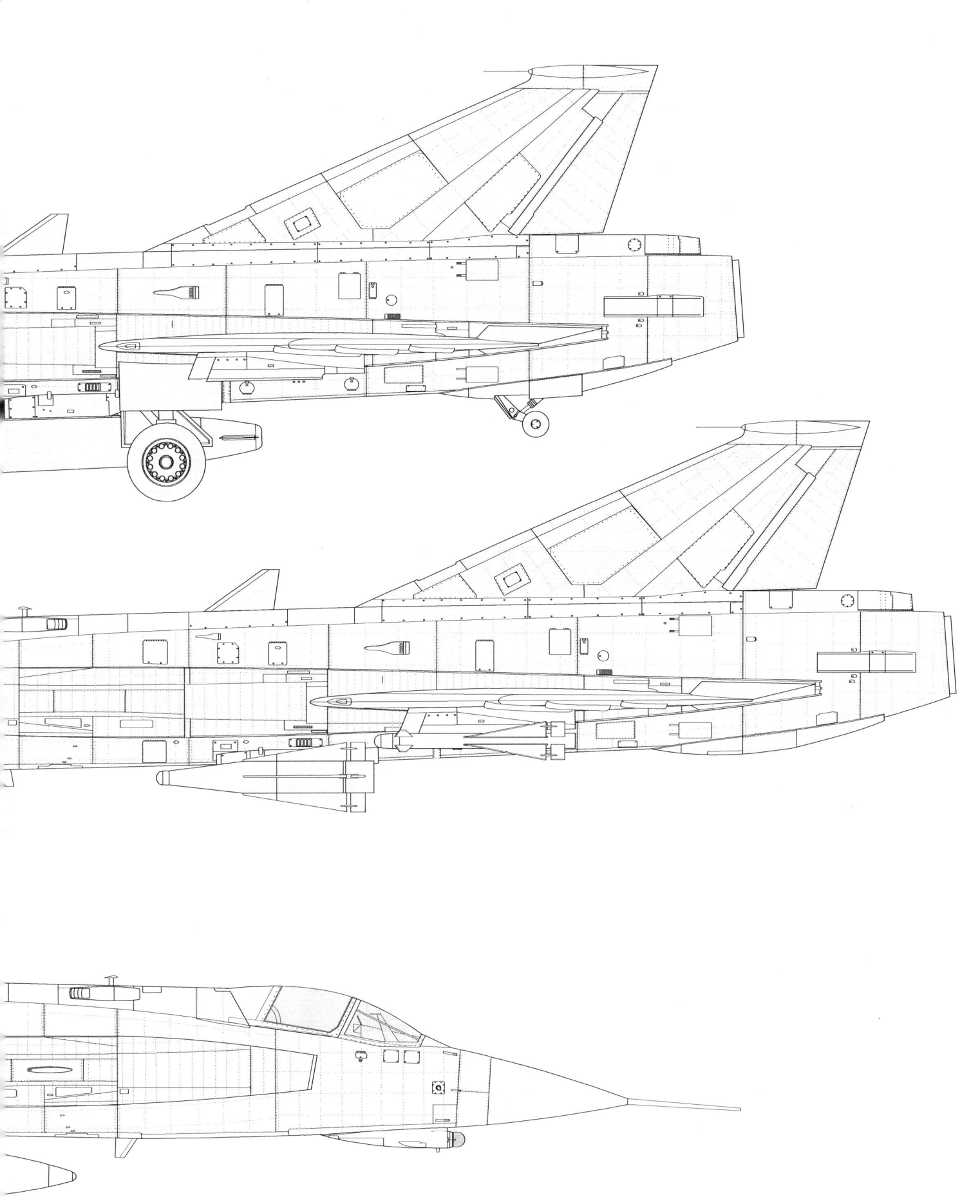

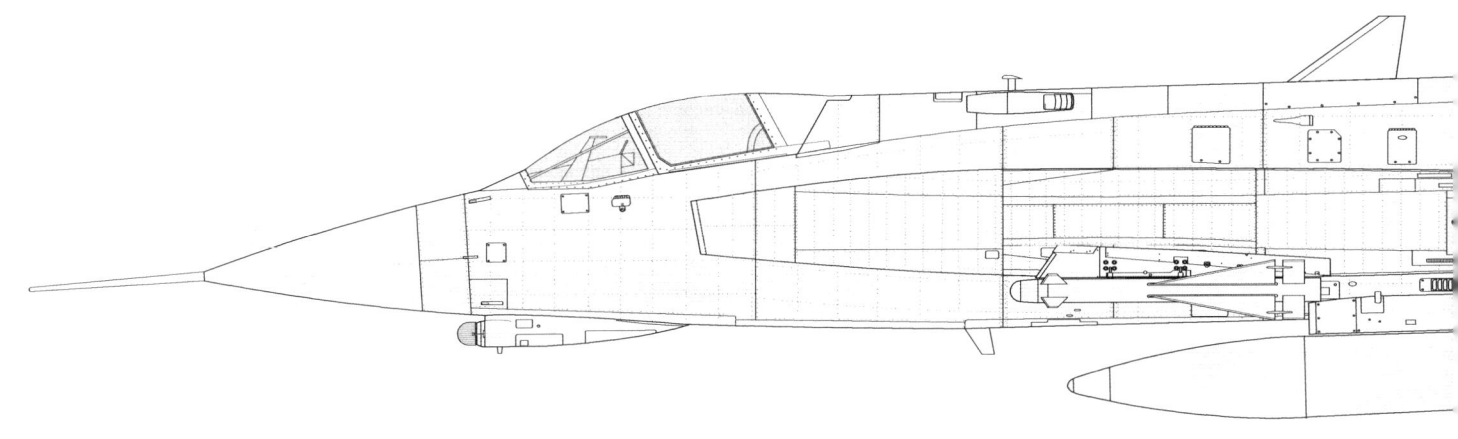

J 35J

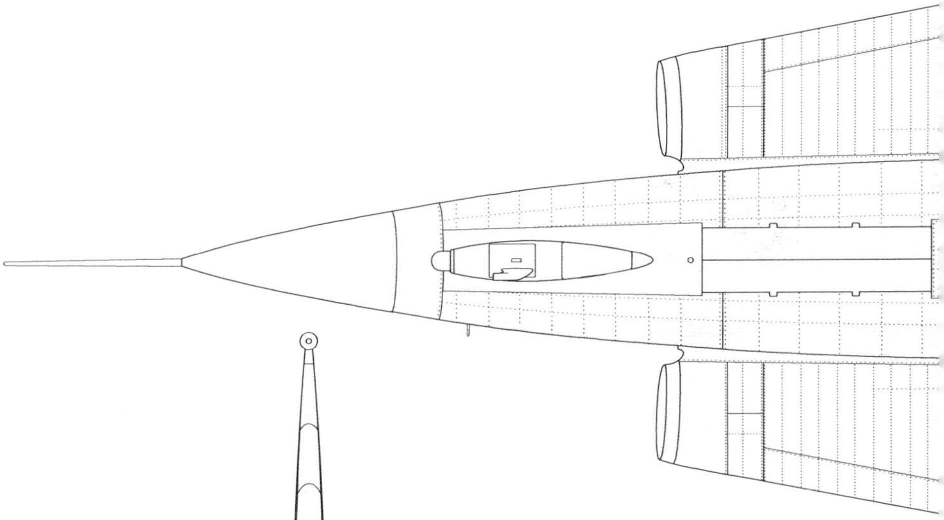

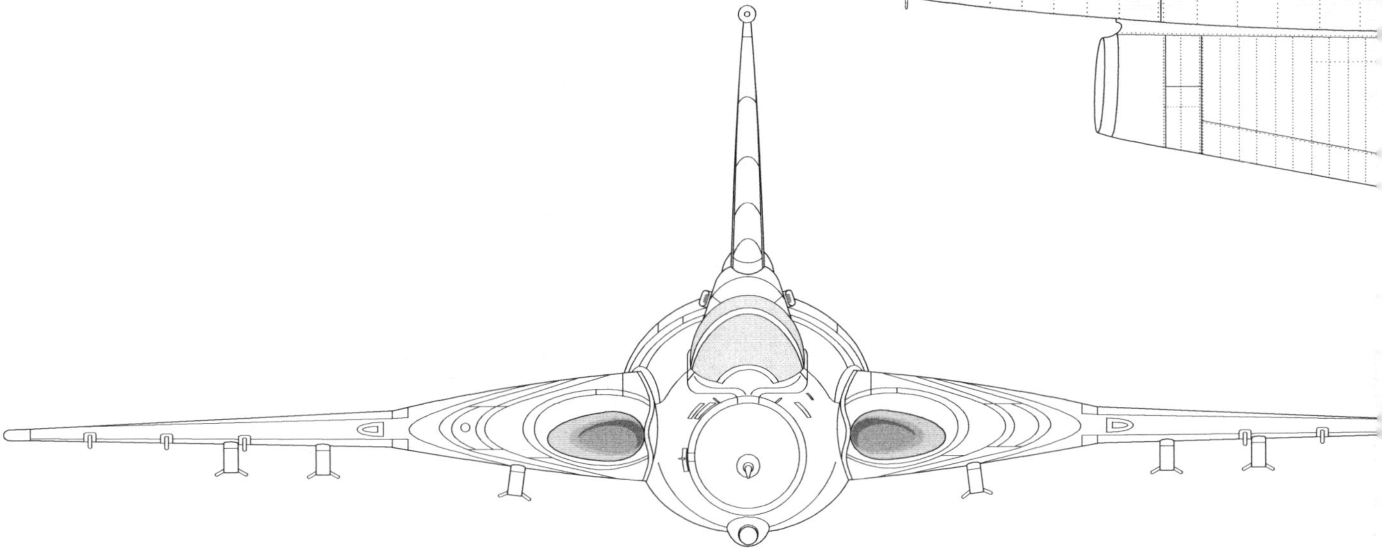

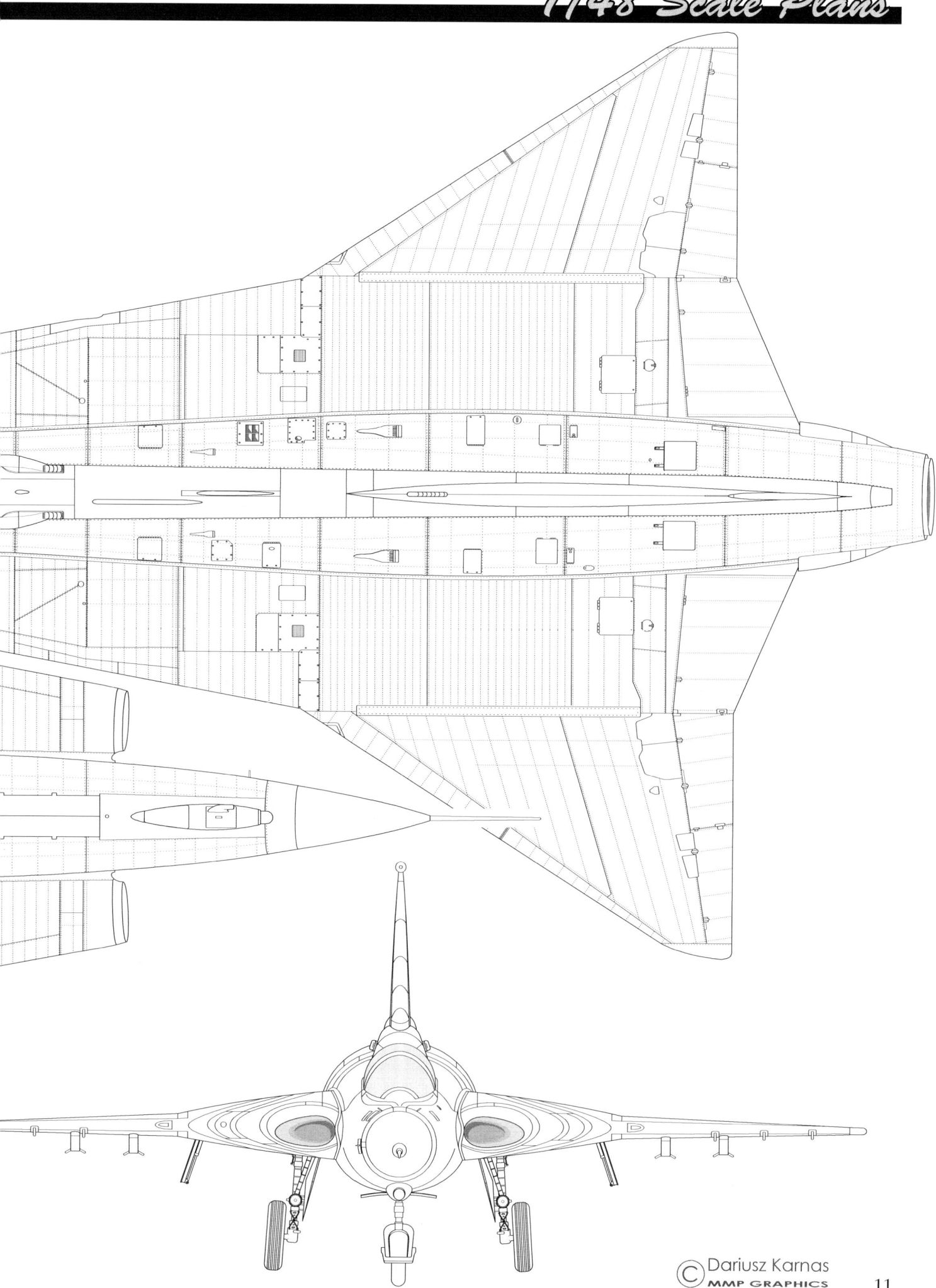

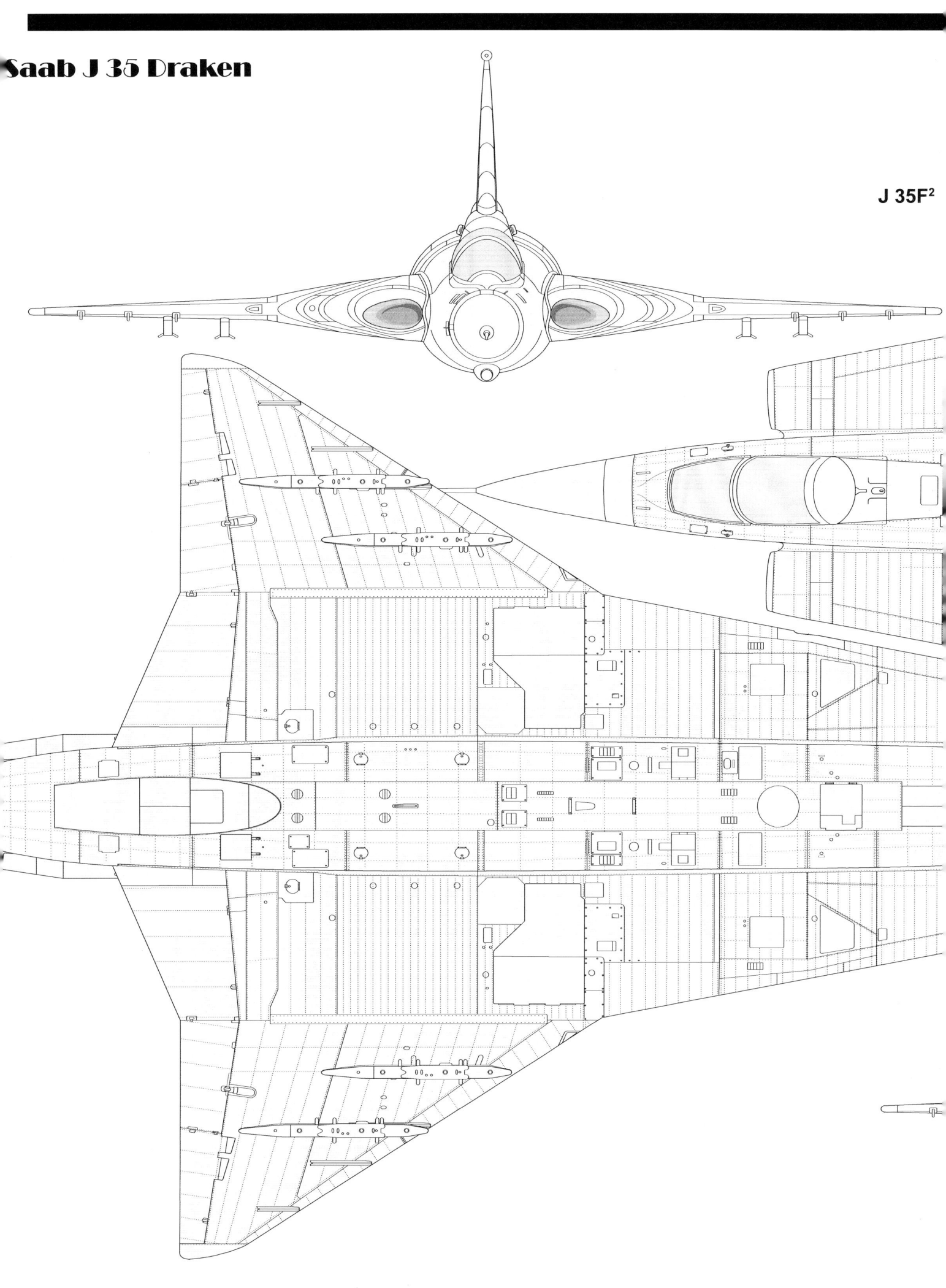

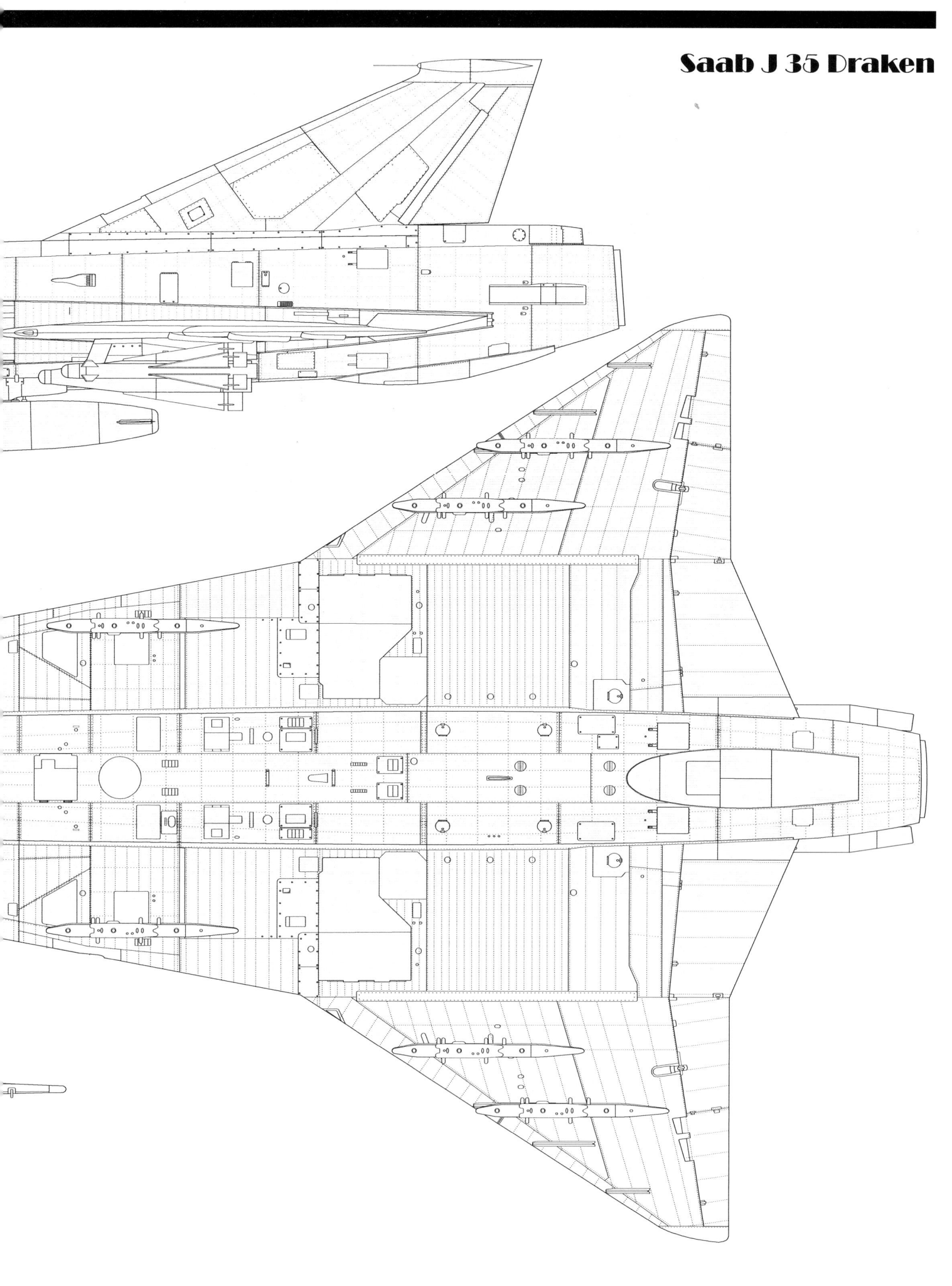